THE FIRST 5 FUCK$...To Give
The Brutal Truth About The Money You're Owed

Published by Epic Author Publishing | EpicAuthor.com
EPIC AUTHOR

General Disclaimer: This book isn't legal advice. It isn't medical advice. It isn't financial advice. It's real-world advice — battle-tested and experience-backed — from someone who's fought the system and won. I'm not a lawyer. I'm not a doctor. I'm not a plumber. I'm a consultant who knows where your money is — and how to help you get it fast, legally, and without getting screwed.

VA Claims Disclaimer: The Fulfilled Veteran Project is NOT a Veterans Service Organization (VSO), an attorney, an accredited claims agent, or affiliated with the VA in any way. We do NOT prepare, present, or prosecute VA disability claims. This book is for informational and marketing purposes only — not official government guidance. If you need legal or medical advice, hire a professional. If you need a battle plan to reclaim what's already yours — you're in the right place.

Paperback ISBN: 978-1-947276-71-0
E-book ISBN: 978-1-947276-70-3

First Edition: 2025
Printed in the United States of America.

Please share this with someone who served. Pay it forward.

FOR MILITARY, VETERANS & FAMILIES

THE FIRST 5 FUCK$ TO GIVE

THE BRUTAL TRUTH ABOUT THE MONEY YOU'RE OWED.

THIS IS YOUR SURVIVAL GUIDE AND CHECKLIST TO TAKE BACK CONTROL - FAST

Albert Winks, SDVOSB
Combat Veteran & Advocate for Warriors

BONUS Fuck$

In order to get your *Shit Together*, you first need to get your *Kit Together*.

Do these ASAP:

1. Take The VA Missing Money Quiz [3-min]

Discover The Life-Changing Money & Benefits You Earned & Haven't Received

TheFulfilledVeteranProject.com/compensation-compass

2. Download The Fulfilled Veteran's "Top 10 Fuck$ To Do TODAY"

TheFulfilledVeteranProject.com

3. Connect To The Five Fs Nation

Join us so Facebook and LinkedIn, so we can do this together.

TheFulfilledVeteranProject.com

If you're easily offended, this book isn't for you.

But if you're serious about getting what you're owed—or helping someone you love, take control of their life—then keep reading.

This is your survival guide.

Table of Contents

THE GOVERNMENT OWES YOU MORE THAN YOU THINK

Introduction

You're Getting Screwed.
And It's Costing You Everything.

You served.
You sacrificed blood, sweat, and years.

And now?
You're getting screwed.

Right now, someone else is living off the money that belongs in *your* bank account.

They're driving *your* truck.
Eating steak dinners with *your* cash.
Vacationing with *your* benefits.

And the worst part?
You're letting them.

Not because you're lazy.
Not because you're weak.
But because no one gave you the playbook.
No one trained you for this battlefield.

So here you are. Waiting. Hoping. Bleeding time...

While the clock keeps ticking... your future keeps shrinking.

There's no back pay on lost time.
No second chances on missed opportunities.

But there's still time to flip the script.

You don't have to fight this battle alone.

The Invisible Fight ...It's more than the money

Why is money section buried in Chapter 3?

Because those of us who've worn the uniform know:
Service was never about the money.

One of my Mentors—who helped me begin my VA journey —received his 100%. Then, shortly after his daughter's wedding, he took his own life. One dark attempt with a struggling Veteran hit him hard—and he didn't recover.

That's why this book starts with **faith, purpose, and people**—not payouts.

God is first.
He gave us Abundance. And He gave us Free Will.

If you're ever in a dark place—**reach out**.
Talk to your God. Talk to your fire team. You are never alone.

He was buried with a folded flag and unanswered questions.

Because money doesn't fix what the system steals from you.

Because this fight isn't just about a payout — it's about taking your life back before it's too late.

If you think you're the exception, **you're wrong**.

If you think waiting will fix it, **you're wrong**.

If you think the system cares about you, **you're dead wrong**.

They don't want to *pay you*.

They want to *delay you.*

They want you to bleed out quietly and disappear.

I wrote this book because silence is the enemy.

I wrote this book because action saves lives.

I wrote this book because YOU are worth fighting for — even when nobody else is fighting for you.

This isn't just about money.

It's about **Faith**.

It's about **Freedom**.

It's about **Fulfillment**.

It's about making sure the next funeral isn't yours.

If you're tired of waiting, hoping, and bleeding...

If you're ready to fight the real war for your life...

You're in the right place.

This book is your bird off the X.

This is your higher level of care.

This is your mission brief for survival.

The Silent War is real.

And it ends when you choose to stand up and fight back.

Let's move.

The System Isn't Built to Pay You–It's Built to Delay You

Since leaving the Army, I've been blessed to help deliver over **$10 Billion** in payouts to Veterans—the best and deadliest humans on earth.

Not because I'm a magician.

Because I understand the evil system—and I've guided you guys through packets, pathways, and the teams to win.

The government hides your money in different places. Each one requires a different packet. A different form. A different fight.

They count on you not knowing that. **They make it complicated on purpose.**

But I know where to look. I know what to ask.

And I know how to get what's yours—**fast**.

Maybe you're thinking:

"This book probably isn't for me. I'm not a Navy SEAL or MARSOC guy."

Wrong.

Every senior leader I've ever worked with—from W4s to O6s—says the same thing:

"If this is hard for us, Joe is fucked."

They're right. That's why I wrote this book.

Because **time is your enemy.** And every day you wait, they win.

The government isn't your friend.
The VA isn't your friend.
The banks aren't your friend.
The system doesn't care if you bleed out.

It's designed to **stall**, **delay**, and **wear you down** until you quit—or die.

Every Veteran who gives up is one less payout they have to make.

But you're still here.
And it's time to fight for your Fulfillment.

The Real Numbers Are Brutal

Over **5,000** Military and Veterans search for a VA calculator **every single day.**

That's over 5,000 daily cries for help. And nobody's picking up.

And still—**22 Veterans commit suicide every day**.

You've probably heard that stat before. But have you ever asked *why*?

It's because we're ignoring the 5,000 asking for help... every day.

It's because the system is **failing them—on purpose.**

And that tells me four things:

1. You're lost.
2. You're alone.

3. You don't know your next steps.
4. You don't know who to turn to.

The three most common truths we hear from Military, Veterans, and Families:

1. "I don't know what I don't know."
2. "I wish I'd known this sooner."
3. "I wish I'd done something sooner."

This book eliminates those regrets.

The Five Force Multipliers

This book isn't just about getting your money.

It's about **stacking the First Five Force Multipliers** that make you lethal again—on your terms.

- 🔥 **Faith** – Find your mission. Without purpose, you drift.
- 🪖 **Foxhole** – Build an inner circle that strengthens you—not drains you.
- 💰 **Finances** – Get your money. Keep your money. Make your money work for you.
- 💪 **Fitness** – You're not in the fight if you're broken. Health = wealth.
- 🛡️ **Freedom** – Time, money, mission. Own them all. No compromises.

These aren't "nice to haves." They're weapons.

And they only work if you take action.

No one is coming to save you.
Not the government.
Not your old CO.
Damn sure not little *g* gov't or little *g* google.

You have two choices:

1. **Keep waiting. Let them bleed you dry.**
2. **Or get in the fight. Stack your force multipliers. Take control of what's yours.**

You already know what needs to be done.
Now you just need to *claim* it.

This Book Is Your Bird Off The X

This book is the bird that gets you off the X.
It's the higher level of care you've been waiting for.
It's the checklist. The mission brief. The battle plan for what comes next.

Military and Veterans...especially your Families don't know where to start.

Most wait too long.

Most lose massive amounts of time, energy, and wealth they'll never get back.

Not you.
Not this time.

A Final Word Before We Begin [Asshole to Advocate]

Let me tell you about one of my Clients—a Marine Colonel (O-6).

We worked together multiple times, but he struggled to trust the process.

He'd start strong. Then stall.

He'd check his status updates obsessively.

Get frustrated. Jaded. Angry.

I kept telling him the same thing I'll tell you:

"Trust the process."

Most of my Clients get paid before they even get a letter. But he kept watching the mailbox. Obsessing. Doubting.

Then one day, I get a text:

"HOLY SHHHHIIIIIIITTTTT. IT HAPPENED."

I called him.

The man was sobbing.

Not from weakness—but from relief.

He had fought the system for years.

And he finally won.

It's Your Turn

What you do with the money is up to you.
But first—we've got to get it in your hands.

This is your survival guide.
This is your mission brief.
This is the bird.

You've popped smoke. I'm inbound.
Let's get you out of here.

THEY DON'T WANT TO PAY YOU. THEY WANT TO DELAY YOU. THEY'D RATHER LET YOU BLEED OUT THAN CUT A CHECK.

I

THE FIVE FORCE MULTIPLIERS

Winning On Your Terms

(Stop hoping. Stop waiting.
Here's how you TAKE what's yours.)

1

FAITH

Your Immovable Foundation

"I have come that they may have life,
and have it to the full."
—John 10:10

I'm in one of those underwater caves you see in places like Hawaii.

No goggles. No wetsuit. No tanks. No flippers. Just me in my swim shorts.

I come up for air in a small pocket—just enough to get my head above water. The cave is nothing but rock, darkness, and silence. I don't know the way out. I don't even know which direction to go.

Completely disoriented.

There's no north, south, east, or west when you're trapped in an underwater cave. No markers. No signs. Just a gut feeling. And in that moment, I realized something—no amount of arrogance, pride, or so-called toughness was going to get me out.

So, who do you turn to when there's no way out?

Who did *I* turn to?

My Heavenly Father.

Every war, every battle, every struggle starts the same way—on my knees. *Help me.* And just like the Bible says, my Father showed up.

It wasn't a voice. It wasn't a map. It was a presence. A deep knowing. It was as if He placed His hands over my eyes and mouth and whispered, *I've got you. This is going to be uncomfortable. But trust Me—you will be better on the other side.*

So, I took a deep breath, closed my eyes, and started paddling—blindly, but with faith.

And then... I emerged.

Now, let me be clear—I wasn't really in a cave.

But I *was* drowning.

I was drowning in a marriage that was over.

Looking back, I can see it now—my Father knew I would never leave. I loved my kids too much. I convinced myself things weren't *that* bad. I put their well-being ahead of my own Fulfillment. But deep down, I knew. I was becoming less of myself every day.

In the military, we do After-Action Reviews (AARs) to assess what went wrong and how to improve. I should have been doing the same in my marriage—taking a hard, honest look at where we were headed. But I didn't. And over time, I found myself lost, just as I was in that cave.

What I didn't know then was that my Father was already working. He was protecting me. He was preparing me. There were things I wasn't ready to see, truths I wasn't ready to face. If I had learned them sooner, the process would have been even more painful.

But faith is the first step.

Faith is the ultimate weapon.

I went to Afghanistan knowing I was protected. And we all came back. That's why I believe. That's why I'm here. To hear you. To help you. To stand with you.

Evil only has power when good people allow it. And I refuse to give it power.

You've been through struggles, too. Or if you haven't yet, you will. That's life. And when that moment comes, don't tell God how big your problems are. Tell your problems how big your God is.

When you pursue your purpose, obstacles will come. When you start to drift off track—when the weight of responsibility pulls you away from what you were *meant* to do—realign. Delegate what drains you. Focus on what fuels you. **That's faith in action—trusting that your path has already been cleared.**

Think of a skier navigating through a dense forest. The untrained eye sees nothing but trees—obstacles everywhere. But the experienced skier? They see the clear path. They don't focus on the danger. They focus on the way forward.

That's what faith is. That's what trust looks like. Don't fixate on the obstacles. Stay the course.

Now, post-divorce, I still face challenges.

I still navigate time away from my kids. But I'm okay. *We're* okay. And I now know things I'm grateful I didn't know back then. Because if I had fixated on the wrong things—if I had stared too long at the "tree" in front of me—I might have crashed.

That's what my Father was doing all along.

He was saying, *You got yourself here. But I have a plan. Hold your breath. This is going to hurt. But you will be better on the other side.*

And I am.

And so will you.

Our Heavenly Father is the ultimate weapon.

Living Faith Daily

Faith isn't just something we talk about—it's something we live. And our Heavenly Father doesn't just *want* us to follow His plan; He *designed* us for it. But sometimes, that path isn't immediately clear.

So here's what I do every day, **I take a breath**.

Every Breath Is Worship

I believe that every breath we take is a silent *I love You* to God.

Just as we parents find joy in simply watching our children grow, our Father finds joy in *us*.

That's why I start every morning the same way—with intention.

My Simple, Powerful Prayer

When I wake up, I say a simple prayer:

"Father, thank You for yesterday. Thank You for today. Put me where You need me, and give me the strength to handle it."

Later in the morning, when I sit down to breakfast with my daughters, we hold hands and pray:

"Father, bless this food to nourish our bodies. Give us strength to bring joy to others. Amen."

It's not complicated. But it's powerful.

Because faith doesn't have to be complicated.

Faith Is Action

I'm not a biblical scholar. I don't have all the answers. But I know this—**one of the greatest ways we show God we love Him is by *bringing joy to others***.

And that's what I strive to do.

Not just surface-level happiness. Not fleeting moments of excitement. But real, deep, intentional joy. The kind that uplifts, strengthens, and fuels purpose.

I know I'm not perfect. None of us are. But the goal isn't perfection—it's progress. It's living each day better than the last.

The Hard Truth About Purpose

"So do not fear, for I am with you;
do not be dismayed, for I am your God.
I will strengthen you and help you;
I will uphold you with my righteous right hand."
— Isaiah 41:10 (NIV)

Everything is through Him and for Him.

I heard a great line the other day:

"It's easier for our Father to tame a mustang than to motivate a mule."

Our Father has me on His mission to help you to your Abundance.

I'll be honest—what I do is not easy.

Nobody *wants* to stand up and say the hard things. Nobody *wants* to fight battles they never asked for. And honestly? I never *wanted* to write these books.

I wanted to live a quiet, simple life.

But I can't.

Because I know what I know. And I *can't* sit back while people—good people—stay stuck, unfulfilled, and missing out on the Abundance our Father *intended* for them.

I won't stay silent.

Because I believe this with everything in me:

Stop telling God how big your problems are.
Start telling your problems how big your God is.

That shift changes everything.

Faith Makes You Unstoppable

Bringing joy to others fires me up. It gives me energy. It strengthens my faith. It makes my journey intentional.

And in the next chapter, we'll talk about shaking off what's *unnecessary*—the weight that slows us down and holds us back.

Because here's what I know:

My Father is bigger than my problems.

And He's bigger than yours, too.

2

FOXHOLE

Choose Your Circle Wisely

Choosing Your Team and Living with Purpose

One of life's hardest truths is that while you get to choose your team—your family, friends, and network—not everyone you surround yourself with truly has your back. Some people will hold you back. Others will lift you up. **The key is knowing who belongs in your inner circle.**

I've spoken to men who have gone through multiple divorces, lost half of everything, and lost custody of their kids. I get it. I've been divorced twice myself, with children from both marriages. It's exhausting to build something meaningful only to watch someone else take it away. If I had a time machine, **I'd go back and tell my younger self to pay closer attention, to make different choices, to recognize the warning signs earlier.**

Rethinking "Happy Wife, Happy Life"

"Happy wife, happy life" might rhyme, but it's not wisdom. I now see it as a dangerous mindset. It puts men in a passive position, sacrificing their own happiness and purpose. Instead, ask yourself:

What does my perfect day look like?

For me, it's about harmony. In the morning, I want to be present with my kids, make them breakfast, take them to school, and then focus on my work with purpose. I want time for myself, time with God, time with my children, and time with my partner. Life isn't about chasing status symbols or external validation. Money isn't the goal—it's just a tool.

At the end of the day, we only get today. And if we can bring joy to others while living with integrity, that's what truly matters.

The Power of the Right People

You need the right people in your foxhole—people who truly support you. In Japanese business culture, there's a practice where before making a big decision, at least one person in the room is required to play the role of the skeptic. This forces the group to consider different angles and make sure they're making the best choice.

You need people in your life who challenge you in a constructive way, who hold you accountable, and who push you to think differently. If they're not adding value—cut them loose.

Dealing with Haters and Toxic People

The higher you climb, the more people will try to tear you down. Some are just trolls, others are deeply hurt individuals projecting their pain onto you. It's not your job to fix them. Your job is to stay focused on your own path.

It's like the analogy of skiing I mentioned. If you focus on the trees, you're going to hit them. Instead, focus on the clear path ahead. Let the obstacles be someone else's problem.

Some arguments are pointless—like wrestling with a pig. The pig enjoys the fight, and you just end up covered in mud. I give people the benefit of the doubt, but once I see they're not worth my time, I move on.

Trust and Performance

When building a team—whether in business, relationships, or friendships—you need both **trust and performance**.

- A high-trust, low-performance person? You'll lose faith in them quickly.
- A high-performance, low-trust person? That's toxic. You can't have that either.

The best teams, whether it's Navy SEALs or high-performing businesses, run on both. And in the civilian world, unlike the military, you get to **choose** your team.

Separating Work from Life

As an entrepreneur, my work and life blend together. But I've learned how to shut it off. When my workday ends, I close my laptop and shift my focus to my kids.

Michael Keaton once said, *"You'll never regret spending time with your kids."* That stuck with me.

When my kids ask me to pick them up or spend time with them, I ask myself, *Why not?* Instead of scrolling on my phone or getting lost in emails, I put them down and show them they matter.

Leadership and Ownership

In business, I operate by a simple framework: **Sales, Marketing, and Management:**

- **Sales**: That's on me. If my bank account is low, it's my responsibility.
- **Marketing**: Their job is to put people on my calendar. If I look at my schedule and it's empty, I know where the issue is.
- **Management**: They handle the behind-the-scenes operations. I don't need to micromanage, just ensure things are running smoothly.

I give my team ownership of their roles. I tell them: *Make my idea better. Be creative, but be realistic. Give me something that works.*

Building a Stronger Future

Choosing the right people—whether in business, relationships, or friendships—makes all the difference. A lawyer once told me, *"An ounce of prevention is worth a pound of flesh."*

If a simple conversation about getting a pet turns into a major argument, what do you think will happen when real challenges arise?

Choose wisely.

I've been through two divorces, and I'm better off now because I've learned who I am and what I want. I don't waste time on the wrong people anymore. I focus on my kids, my work, and surrounding myself with people who align with my values.

At the end of the day, it's about **faith, purpose, and taking control of your life**. Find your people. Set your standards. And don't settle for less than what truly Fulfills you.

What Is Love?

I went on a date recently with a woman who was sharper than most. She asked me a simple but powerful question:

"What is love to you?"

I started talking about my kids—how I love them, how I hold them, how they mean everything to me. I went on too long, over-explaining, until I realized I was rambling. Then I pivoted, trying to guess what she really meant—something we all do too often in conversations.

Finally, I stopped myself and said, *"You know what? Let me turn that question back on you. What is love to you?"*

She gave me an answer that stuck with me.

"Love is handing someone your heart and trusting them not to break it."

She explained further,

"If you need to nurture it, do so. If you need to apply a little pressure, communicate with me. Help me alleviate the weight. But don't break my heart. Don't crush me. Don't betray me. Don't string me along. Love is trust."

That hit me hard.

We overcomplicate love, don't we? We try to define it with grand gestures, deep philosophies, or long-winded explanations. But really, it's simple: love is trust. It's placing your heart in someone's hands and knowing they won't shatter it.

Pigs, Trolls, and the Art of Distraction

There's a saying: **"Never wrestle with a pig. You'll both get dirty, but the pig likes it."**

Some people thrive in the mud. Their entire purpose is to pull you down.

Trolls. Haters. Online bots. Noise.

I've fallen into that trap before—let negativity get to me, let someone push my buttons. And I'll be real with you: I've done things I'm not proud of because of it.

I once had a day where everything piled up—stress, frustration, some troll getting under my skin. I lost it.

I slammed a chair down. It shattered into pieces.

Even a year later, after moving out of that house, we were still finding shards of that chair. It was a reminder of that moment—just how easy it is to let emotions control you.

And I was embarrassed. Because losing control like that? It's childish.

"When I was a child, I thought like a child. But now that I am a man, I must act like one."

We all have breaking points. But we don't have to let people push us to them.

Own Your Choices

If you keep stacking the deck against yourself, you shouldn't be surprised when you lose.

I've always been a positive guy. But I also know this: we allow negativity into our lives. We surround ourselves with the wrong people, make the wrong choices, and then act surprised when things don't go our way.

Here's the hardest truth I had to accept:

It was my fault.

Taking ownership is tough, but once you do, you take back control. And the crazy thing? Even after I admitted my mistakes and grew from them, people still tried to blame me for things I'd already owned up to.

At some point, you have to stop engaging.

If someone called you up and said, *"Your hair is blue,"* you'd laugh it off, right? Because it's obviously not true.

That's exactly how you should treat baseless accusations and negativity. Let them slide past you. Don't even let them stick.

Stay focused on the path in front of you.

The Right People in Your Foxhole

In skiing, if you focus on the trees, you'll hit them. If you focus on the clear path, you'll glide through.

Life works the same way.

Surround yourself with the right people—the ones who

help you move forward, not the ones pulling you off course. If you keep losing, check your circle. The wrong mentors, the wrong friends, the wrong relationships—they'll sabotage you every time.

If you're constantly making choices that put you in bad situations, how can you expect different results?

You need people who hold you accountable, people who challenge you the right way. Those are the ones you want in your foxhole.

Flipping the Script on Doubters

I used to say, *"Keep me posted on your journey,"* when people told me they were going to do something on their own. But now, I flip the script.

<u>If someone tells me they're ready to go solo, I say:</u>

> *"Awesome! You think you're strong enough? I want to watch. Come sit next to me. Show me how you navigate this process. If you're that capable, I want to learn from you too."*

<u>One of two things will happen:</u>

1. They'll rise to the challenge, and we'll build mutual respect.

2. They'll realize they weren't ready and they'll come back humbled, ready to learn.

I've seen guys leave, thinking they had it all figured out, only to return months or years later, broken, searching for guidance.

The truth is, I don't need to convince anyone to follow my path. I'm just here to offer a way forward for those who want it.

Haters vs. Hustlers

Haters will always talk.

But here's what I've realized: **the difference between a hater and a troll is intent.**

- **Trolls** exist just to bring you down. They don't even believe what they're saying. They just want a reaction.
- **Haters** might actually know something, but their bitterness comes from one thing—they don't know how to be you.

I used to brush it off. Now? I challenge them.

"You sound like a strong person.
Why don't you try to do what I do?
Step into my world. Let's see if you can handle it."

Because most of the time, they can't.

The Six-Month Rule

One of the sharks on *Shark Tank* has a rule: every new hire gets a six-month trial. No matter who they are, no one is permanent before then.

I love that.

Six months is enough time to see someone's true colors.

I've had people try to game the system. Someone once asked me, *"Hey man, can you put me on your payroll so I can get approved for a mortgage?"*

Absolutely not.

I get it—you want to buy a house. But you have to **earn** it. You can't fake your way to success.

Be righteous. Do things the right way. And don't let people into your foxhole who don't belong there.

Final Thoughts

At the end of the day, it all comes down to:

- **Knowing what love really is**—trust, care, and not breaking the people who trust you.
- **Avoiding the pigs and distractions**—not getting pulled into the mud by trolls and negativity.
- **Taking ownership**—recognizing that if your life is off track, you have the power to fix it.
- **Surrounding yourself with the right people**—because the wrong ones will sabotage you every time.
- **Flipping the script on doubters**—challenging them to step up instead of wasting energy proving yourself to them.

You can't control everything. But you can control your mindset, your reactions, and the people you choose to keep close.

And that makes all the difference.

Turn the page...

3

FINANCES

Get the Money They Owe You

THE MONEY YOU'VE EARNED—HOW TO GET IT FAST

Imagine 5,000 American civilians calling 911 each day. On the other end, the phone just rings—no one answers. No voicemail, no assistance, just ring, ring, ring.
No Hope

If that were to happen, the backlash would be immediate, and the issue would be unfucked fast. The world would not tolerate such a failure.

But here's the harsh reality: this isn't happening to civilians. It's happening to our Military and Veterans. Each day, 5,000 Veterans search online for a "VA calculator."

That's not just a statistic; it's a cry for help. It tells me four things about these individuals: they're lost, they're alone, they don't know their next steps, and they don't know who to turn to.

I'm a problem-solver by nature, and while I'll provide solutions, I also need to lay the groundwork. I want to explain why this book exists and how it's designed to help.

Two Types of Clients

I've worked with two main types of Clients:

The Young – For those at the start of their Military journey, my goal is to help them be **whole** from the beginning of their service. These are lessons you and I didn't have the chance to learn early on.

The Old – For those who have already served, the mission is to make things **right**. It's about helping them get back on track and find fulfillment moving forward.

Faith and Purpose

The first chapter of this book is about faith, and there's a reason for that. Recently, I heard a saying: "It's easier for God to tame a mustang than to motivate a mule."

That line hit me deeply for three reasons.

First, I grew up on a cattle ranch in Texas, so the metaphor resonated.

Second, in the Army, I was a Mustang—a term for those who transition from enlisted ranks to officers. This felt like God speaking to me in His own unique way, not through words but through vibrations, feelings, and purpose.

Thirdly, I used to drive a mustang. I feel like God has taken the wheel of this mustang and is doing donuts in the parking lot of your life...to get your immediate attention to aim you towards your Abundance

Let me be clear: I don't enjoy being in front of a camera or writing about tough topics...telling you bad news. But this is my calling. This is God saying, "Albert, you need to get out there and help these men and women find their Abundance. Evil is stealing from them."

The Battle Between Good and Evil

Good and evil are real. Evil doesn't create; it only takes. And who does evil target? Good people who either don't know better or don't do better.

That's why I'm here—to lay the groundwork, identify the problems, and offer solutions. Evil is stealing from our Veterans, our widows, and our families. It's time to stop that theft and reclaim the Abundance and Fulfillment God intended for us.

Here's an example: widows of Veterans are **never** paid what they're owed. Why? Because the system is designed to take advantage of them. That's why I don't charge widows; I charge others so that widows don't have to bear that burden.

The Knowledge Funnel

Many people feel stuck because they don't know what they don't know.

That's one of the top things I hear from clients:

> "I don't know what I don't know."
> "Albert, I wish I had known this sooner."
> "I wish I would have done something sooner."

My job is to guide you through what marketers call the "knowledge funnel." Right now, you're outside of it, unaware of what's possible. My role is to get you inside that funnel so you can start taking control of your life, your finances, and your future.

The Reality of Compensation

The VA uses the term "disability," but that's a lie and scare tactic. The correct term is "compensation and pension." The government recognized years ago that military service takes a toll—on our bodies, our minds, and our families. Compensation and pension were designed to make things right.

But evil does what it does best: it steals. Instead of ensuring Military and Veterans receive what they're owed, the system denies claims, creates confusion, and makes people feel ashamed for asking for what's rightfully theirs.

Here's the truth: there are 20 million living Veterans today, and almost none of them are receiving the $5,000 a month they earned. That's over $100 billion a month being withheld. We must take care of those who take care of us.

Building Fulfillment

My goal is to help every Veteran reach 100% Permanent and Total (P&T) status as quickly and compliantly as possible. This status maximizes both **Your Money** and **Your Freedom**—key components of Fulfillment. It ensures you're compensated for your service while retaining the ability to live fully, whether that means owning firearms, being a pilot, or pursuing other passions.

For younger Service Members, my goal is to start them off right...making them **Whole** from the beginning—educating them on life-changing benefits like the GI Bill, VA home loans, and financial planning. By setting them up for success early, we prevent the gaps and mistakes that so many of us experienced. **Cash Flow**

For older Veterans, it's about making things right and guiding them through the steps to reclaim what's owed to them.

A Call to Action

This book is just the start. I need your help to spread the word. Tell your friends, share the resources, and let others know that there's a path to Abundance and Fulfillment.

Together, we can stop the theft of what's rightfully ours. Together, we can ensure that no Veteran is left behind, ignored, or forgotten. Let's move forward, guided by faith, and reclaim the Abundance that our Heavenly Father intended for us all.

There are 3 Levels of Money:

1. Right Now Money
2. Mid Range for your kids in college
3. Long Range Death Benefits

Let me save you some time:

You're missing THOUSANDS of dollars—every single month.

CHECK YOUR BANK ACCOUNT RIGHT NOW

If you're a non-retiree with a family of four, you should receive at least $5,000 per month, tax-free.

If you're retired or medically retired, you should get $7,500 per month through Combat-Related Special Compensation (CRSC).

Are you getting that?

I didn't think so.

THE MONEY THEY NEVER TOLD YOU ABOUT

Beyond your GI Bill and VA pay, there's an entire world of hidden benefits that most Veterans never access:

- SMC-S—Ever heard of it? You should have.
- Property tax exemptions— States will reduce or eliminate your property taxes
- Education benefits for you—Student Loans wiped out.

- Education benefits for your kids—
- College tuition and fees reduced/removed.
- Ch 35 monthly Stipend
- Additional compensation streams—Some of you could double what you're receiving.

Why weren't you told?

Because they don't want you to know.

Evil's plan is simple:

Bury you in paperwork.
Make the process confusing as hell.
Hope you give up and go away...until you die and get off their payroll...then they'll steal from your grieving family

And guess what?

It's working.

HOW MANY VETERANS ARE GETTING EVERY DOLLAR THEY'RE OWED?

Out of 20 million Veterans and Service Members, how many do you think are actually receiving everything they deserve?

Would it be... zero?

I've been at this for over six years. I've been blessed to help recover over $10 Billion in lost benefits.

And not one single Veteran has ever told me:

"Albert, I'm getting every cent I'm owed, and I couldn't be happier."

Not one.

How many Widows are getting paid All of the money they Should Receive?...None

If Evil is willing to steal from Widows...You're a way easier target for Evil to take from

THE VA'S DIRTY SECRET—AND HOW TO TAKE CONTROL OF YOUR MONEY

The VA makes decisions based on three levels of evidence:

Administrative Evidence—Your paperwork.

Medical Evidence—Your records (which they love to use against you).

Legal Evidence—The game-changer most Veterans never think about.

Most Veterans lose their claims because they don't know how.

Here's the reality:

> Say the wrong thing on your claim? Denied.
> Submit old medical records? Denied.
> Miss one legal loophole? Denied.

That's why I guide my Clients to perfection.

Because the VA is looking for a reason to say no.

100% P&T ISN'T THE END—IT'S JUST THE BEGINNING

Some of you think, *"If I just get to 100% , I'm set."*

Wrong.

Property tax exemptions—If you're not at 100% P&T, you don't qualify.

College tuition waivers for your kids—Again, no 100% P&T? No benefits.

More hidden money—And I mean a lot more.

TIME IS YOUR BIGGEST CURRENCY

Every month you wait is money lost forever.

I sat at 90% for five years and lost over $65,000.

Don't make the same mistake.

START WITH ME—STAY WITH ME

This isn't a one-time thing.

There are 2 terms I use that I'd like to Identify and Clarify for you

Making You **Whole**.

Youngsters just entering Service or in your earlier years of your Service.

Help me make your Fulfillment Journey begin as early as possible.

As a for-profit, I can help wipe out the ability for Evil to ever get its hands on your Abundance.

I'm able to guide you from the very beginning of your Service to show you where GI Bill begins and how to use it for your benefit as quickly as you can

I'll guide you to other earlier benefits like VA Home Loans – How you can begin your Abundance by renting out your earlier houses to begin your cash flow.

If you want great advice...go to rich areas of Florida...sit next to an old guy and let him tell you all about how easy money real estate is...the earlier the better.

I call money...Freedom Tickets.

Would you like the Freedom to stay in and Serve a great career? Great. You have to be told that most folks only Serve 1 or 2 tours thus getting out with not much in your tank on Fulfillment.

Even those who Serve 20 years are still young...with 30, 40, maybe 50+ more years left in life

Ask the "old guys" what they wish they would've done earlier to gain their Freedom Tickets and Fulfillment.

I know because they tell me every day.

This is why I build a Team of Good around you.

Money comes in many forms:

- Spending
- Saving

Different Insurances:

- Life Insurance
- Health Insurance

Keep in mind there are tons of other types of money that require:

- Knowledge of the benefit
- Perfection of process

Delivering the money to you in the best order for you.

Making You **Right**.

This is for Older Service Members and Veterans.

Like me, you're older and most money is in front of you rather than behind. Can Veteran's Rights attorneys help sweep up past mistakes?...Yes, and I'll help guide you to these benefits as quickly as possible.

But, you can't be made Whole...some past money has to stay in the past...just Right from here... getting you back on track to your Abundance and Fulfillment.

Your life changes.
The VA rules change.
New benefits get unlocked.

That's why you need a team in your corner—no one wins this fight alone.

Your first step:

Go to my website and start with **Step One—Pay Yourself First.**

This starts your back-pay clock.

From there, I'll guide you through my Red Smoke Campaign and Triple Campaign, making sure you get every single dollar you're owed.

Right now, 5,000 Military and Veterans every day are searching for help.

That's 5,000 Military and Veterans popping red smoke, asking for help

I see it. I hear it. I'm coming to help.

But I need you to help me help more Military, Veterans, and your Families.

I won't stop until I've gotten every last one of you off this battlefield to a higher level of care

Start with me. Stay with me. Let's get your Freedom Tickets, Your Abundance, and Your Fulfillment

Turn the page...

4

FITNESS

Your Body Is Your Weapon

Fitness, Resilience, and Taking Charge of Your Life

Hey guys, physical fitness has always been important to me. I could start this off with a "victim" story, but that's not my style. Fitness has never really been a challenge for me—I've always put in the effort. And today, I want to talk about that journey, from the ground level as an infantryman to my time as a MEDEVAC pilot.

Early Challenges and Overcoming Obstacles

I was born with medical challenges that required multiple surgeries. As a baby, I was underweight for corrective procedures, so it took years before I could undergo the necessary operations. I've had at least four surgeries that I can recall, and I carry the roadmap of scars on my body. While they were inside me, doctors even decided to remove my appendix since they already had me cracked open.

Despite these challenges, I've never viewed them as weaknesses. I don't start conversations with my medical history because I don't see myself as a victim. I bring it up only to show that obstacles can be overcome with the right mindset. I grew up in Texas, and most people don't know the full story of my background—but sometimes, those details matter.

A Lifetime of Pushing Limits

From a young age, I was involved in sports. I was a competitive gymnast and later coached at the elite level. One of my teammates competed in the 2000 Olympics. When I joined the military, I was already extremely fit, but the experience tested me in new ways.

I arrived in Germany as an infantryman in the winter—deep snow, cold weather, and carrying heavy gear. Coming from Texas, I had no experience with this kind of cold. I didn't know how to layer properly, and I learned the hard way. One time, while rucking in knee-deep snow, wrapped in too many layers and a neck gaiter, I suddenly felt like I was suffocating. I collapsed in the snow—not from exhaustion, but from frustration. It was a lesson in adaptation and mental toughness.

Throughout my military career, I always pushed myself. I wanted to max out every fitness challenge. When I saw the German Armed Forces Proficiency Badge, I thought, *That looks cool. How do I earn it?* I didn't care about the minimum requirements—I aimed to exceed the maximums.

500 Miles on a Bicycle and Hard Lessons

Toward the end of my Army career, I took on a 500-mile bike ride across Indiana for a Veterans group. It was five days of riding—100 miles per day. I had no idea what I was in for.

Day one nearly broke me. The first 100 miles taught me more about my body than any training ever had. I learned the importance of proper fueling and hydration. By the end of that first day, I was completely drained. If there had been one more mile, I might not have made it.

The next challenge was the infamous "Hilly 100" in Indiana. If we had started with that, I might not have finished the ride at all. But by then, I had learned from my mistakes—fueling properly, pacing myself, and pushing through the discomfort.

Lessons from Triathlons and Why I'll Never Be a Navy®SEAL

I also dabbled in triathlons, but swimming was my weak spot. No matter how much I tried, I'd sink like a rock. After a couple of tries, I realized I just didn't enjoy it, so I switched to duathlons instead. I believe in doing what makes you happy—if something isn't enjoyable, why force it?

That's why I was never cut out to be a Navy SEAL. I respect those guys immensely, but I have no desire to be cold, wet, and sandy. You can keep the sugar cookie punishment to yourself—I'm good. I'd rather be hot, gritty, and grimy...on land

Taking Charge of Your Fitness and Life

One of the best pieces of advice I ever got was from ROTC: *When in charge, take charge.* That applies to your fitness, your career, and your life. No one is coming to save you—you have to take control.

After leaving the military, I let myself slip . I drank more than I should have, ate too much, and felt constantly bloated. Then, a wake-up call came when I met a fit, older guy at a Special Operations Convention in Florida. He told me, *I eat once a day. I eat to fuel my body, not to overindulge.*

That conversation stuck with me. I cleaned up my diet, focused on quality food, and started eating with intention. Today, at 44 years old, I weigh 149 pounds—the leanest I've been since my Army days.

Find What Works for You

Your fitness journey should be something you enjoy. I love running outdoors, off the beaten path. I don't do treadmills—I need the open air, the Freedom. Whether it's running, biking, weightlifting, or another activity, do what keeps you motivated.

Remember, *you can't lead, fight, or win if you're broken.* Just because you were once fit doesn't mean you'll stay that way if you get lazy. Take charge of your health. Eat less, move more, and listen to your body.

At the end of the day, your fitness is your responsibility. No one else can do it for you.

Turn the page...

5

FREEDOM

No One Gives It. You Take It.

Freedom Isn't Free

People throw around the phrase "Freedom isn't free" like a cliché, but the truth is—it really isn't. Freedom takes action. Freedom takes a fight. And let me tell you, that fight isn't always clean. It's messy. It's exhausting. It'll leave you bruised and battered, but if you want it badly enough, you'll keep going.

I've heard so many excuses—no, let's call them reasons. Real reasons like, "My house is on fire, I can't focus on anything else right now." That's a reason. But most of the time, what I hear are just excuses. And look, I'm not some hardcore Navy SEAL barking orders at you, but I *feel* it—I feel it in my bones—**Freedom isn't free**.

And I'm not talking about going to war with a gun in your hand. There are other battlefields. The battlefield is in your home. It's in your job. It's in your everyday life. And most importantly—it's in your head.

I never really thought about this before, but I heard someone say: *Even the devil can get inside your head.* If that's true, then you better believe other people can too.

If you hear a thought that doesn't sound like you, maybe it's not. So why believe it? Why let it define you? Slow down. Filter it. Push it out. Treat it like someone telling you, "Your hair is blue." If it's not true, it doesn't stick. If it's not you, roll it off.

And every single day, go after that inch. Then that mile.

Fighting for My Son

I told my son this once.

The same son I now have custody of. It wasn't easy. It wasn't quick. It took a year and a half—not of a custody *battle*, but a fight nonetheless. Because the truth is, I didn't battle for him—his mom battled *against* him. Or, more clearly, against *me*. And if I'm being brutally honest—against *herself*.

I remember the day after court, standing in his school.

He didn't know we were coming.

And through the office glass, I saw him.

My son.

He looked miserable as he walked between classes. He looked like his soul had been ripped from him...poor kid.

You ever notice how in every superhero story, the heroes are always outnumbered? There are a thousand mindless minions doing the villain's dirty work. But the hero? The hero usually stands alone.

And some people—some people just *want* to watch the world burn.

I've seen it firsthand.

The system was rigged against me. And sure, I'm not perfect.

I didn't do everything right. But when I tried to be there for my kids, tried to be a father, there were forces at play that did everything in their power to stop me. Not putting them on a plane. Not answering the phone. Not letting them talk to me.

Liam was old enough to make his own choice. We thought it would be simple. Go to the judge, say what he wanted, and that would be that. But that's not how it played out.

Instead of a simple process, it turned into catapults and arrows flying at me from every direction.

Instead of fairness, we got endless court hearings, delays, and accusations.

And six months in, it got worse. They tried to weaponize everything against me. Every word I said was twisted, every moment scrutinized, and at one point, they tried to take *everything*. Not just Liam—but my rights as a father, my Freedom as a man.

They wanted me in jail.

All because I spoke the truth.

And yet, through all of it, I stood my ground. Because I had *Faith*. Because I knew—no matter how much they tried to break me—I would not stop fighting for my son.

And in the end, after a year and a half of hell, he was finally home.

The Cost of Freedom

If you think Freedom is easy, stand next to me for a day. Walk with me for a week. See if you can handle what I go through.

But you don't have to face it alone.

I'll walk with you—shoulder to shoulder—through your storm, through your fire, through your fight.

Because I've been there.

And if no one else shows up for you—I will.

And I keep showing up—because it's not just me doing the work.

It's God working through me.

Through my Heavenly Father, I am able to help others find their own Freedom.

But let me be clear—**Freedom isn't something you're given. It's something you take.**

And it sure as hell isn't *free.*

It costs you *everything*.

But in the end—it's worth it.

II

BONUS FUCK$

The Extra Chapters You Didn't Know You Needed

6

FAST ACTION

Speed Is Your X-Factor

Trust, Performance, and the Path to Abundance

How do you think I've gained the respect, admiration, and loyalty of the deadliest humans on earth?

> Is it my bank account?
> My beautiful good looks?
> The fact that I'm six feet tall and bulletproof?

Not exactly...

So how do you win? How will you *gain* your definition of respect and trust?

Let's sit on that word for a second—**gain**.

Earning Trust, Not Titles

How did I gain the trust of Warriors who could tear me apart with their bare hands?

I didn't go to Ranger School.

I don't have a short tab.

I don't have a long tab. I was told **no** for both Ranger School and SF.

So I pivoted.

I did what I wanted to do.

And because of that, I get to work with the best—every single day.

And I tell them this:

It's **not** about the money.

$10 Billion, that I've helped deliver in payouts to the Veterans I've served. **But that's nothing.** There's so much money out there, more than we can comprehend.

Want to see real money? Walk through the halls of government. Step inside aviation. Visit Rome. Go to the Vatican.

That's **decadence**.

Now, look at me. I'm in what I call my PJs—shorts with a belt, a zip-up hoodie. Comfortable, every day.

Because **none of this is about appearances**.

It's about what I do **every single day** that earns me something more valuable than money—**trust and performance**.

Why They Come to Me

These men and women have been through war, through trauma, through hell. And to the government holding their money, none of that matters. To the system, it's just paperwork.

So why do they come to me?

Trust. Performance.

They know when I show up, I bring truth. I bring hard conversations.

But we don't argue—we **spar**. We sharpen each other. We grow together.

I never tell someone, *"Do nothing."*

Never.

Instead, I pull them in closer and say, *"I love you, but you're overthinking this. Or maybe you're underthinking it. Let's get clear on what needs to happen."*

Because in the real world, sometimes **perfection matters**.

When you're landing a plane, your calls have to be perfect.

When you're making a split-second decision, **speed matters**.

Move too slow? You're already lost.

Cut the Anchor Holding You Down

I see it in your face—even if you don't say it.

You want something. And you *know* you're the one standing in your way.

You're a rocket, **loaded with fuel**, ready to launch.

But you're still tied to the tower.

That tower—whatever it is—**is your excuse**.

And when you read this, when the hair on your arms stands up, when your heart beats faster—that's your body telling you something.

Listen to it.

The thing you want? It's already yours.

Go get it.

Go get it **right now**.

Because Abundance has already been promised to you.

Not easy Abundance.
Not lazy Abundance.
Not *someday* Abundance.

It should have been here already.

But maybe—just maybe—you had to go through what you did to appreciate it.

And that's fair.

Move Faster

Trust and performance.

People love to say, *"Slow is smooth, smooth is fast."*

I get it.

But I work to **standard**, not time.

And time? Time should always move faster.

Some things take longer because they're out of my control. I can't move the moon and stars. I can't flip the calendar forward.

But what I *can* do is **set my vision**.

> I've never been to Greece.
> I've never been to Alaska.

So when do I want to go? What's the best way to get there?

It's not complicated.

We have the internet. Look it up. Make a plan. Execute.

Do Something—Right Now

That's it. That's the entire game.

Do something **today** that moves you forward.

This isn't a long, complicated guide because it doesn't need to be.

All you have is **now**.

Tomorrow isn't promised. Yesterday is already gone.

But right now? **Right now, you have a choice.**

> I'm writing this book *right now*.
> I'm on camera *right now*.
> I'm spending time with my kids *right now*.
> I'm wearing what I want, doing what I want—
> because I built my life that way.

That's my Abundance.

And all you have—**is now**.

Go to my website so you can start with me and stay with me.

Turn the Page...

7

FUEL

Control What Controls You

A lot of people focus on what they eat, but consumption goes beyond food. It's everything you take in—what you watch, listen to, and engage with.

I've always found it strange how some people get so emotionally invested in sports. I was an athlete myself, and even my daughter recently asked me what sport I liked. I had to clarify—was she asking about what I liked to play or watch? When she said "play," I told her football was a lot of fun. Growing up in Texas, football is a way of life, no doubt.

Sports teach valuable lessons—teamwork, discipline, perseverance. But as the Bible says, "When I was a child, I acted like a child. When I became a man, I put away childish things." That comes to mind when I see grown men jumping and screaming at a TV. What kind of void are they trying to fill? What disappointment makes someone cling so desperately to a team's victory or loss?

A lot of it feels intentional—a distraction.

It reminds me of the old Roman strategy: "Give them bread and circuses." Keep people entertained so they don't realize they're being taken advantage of. It's unsettling.

This brings me to my main point—what you consume matters. It's not just food but also the *information you feed your mind.*

I don't let my young daughters spend endless hours on screens because when you give your kids access to the internet, you're also giving the internet access to your kids. That's a terrifying thought.

Michael Keaton once said, "You'll never regret spending time with your children."

Today, I'm taking my girls swimming. After a long winter, it's time to build their confidence in the water. Tough kids are raised by tough parents.

Our Heavenly Father is tough on us too—pushing us to be

better while evil keeps us lazy and complacent. Everything you consume—good or bad—ends up consuming you.

I know this firsthand.

I gained 30 pounds, then lost it. Now, at 148-149 pounds, I've maintained that for nearly two years.

Recently, at an event, a guy recognized me from two years ago and said, "Holy crap, man, you lost a ton of weight!" At 5'6", losing 30 pounds is noticeable. **But it didn't happen overnight.**

People throw around phrases like "fat, dumb, and happy" or "happy wife, happy life." But those are just excuses from people stuck in misery.

If you're feeling that way, I encourage you—eat less, move more, challenge yourself.

There are so many groups out there—running clubs, Team Red, White & Blue, and others. Signing up for something, even a short run, can push you toward a better version of yourself.

Once my book is out, you'll see me out there even more. But for now, I'm focused on helping you help yourself.

Another thing—everything is filled with garbage.

My family and I have even talked about moving to Florida. Indiana has its downsides, and I can't stand seeing chemtrails in the sky. I think a thousand years from now, people will look back and say, "That era was insane—they allowed themselves to be poisoned from the air, land, and sea."

One of the best things I ever did was clean out my pantry and throw away all the junk.

Ironically, my ex-wife went out and bought it all again. I call it "sugar brain"—when your body craves what it's used to, even if it's bad for you.

But here's what changed my life: a guy once told me, "We all eat more than we need to. We're conditioned to eat three meals a day, but we don't need that."

I started eating one meal a day—steak and eggs.

It wasn't immediate, but over time, I lost 30 pounds.

Americans, especially, have fat stores we don't even realize, and our bodies can either burn them the right way or destroy themselves in the process.

At 44 years old, I hadn't seen 140-something pounds since I was 20 and stationed in Germany. And now? I feel better than ever.

One trick I learned—when you feel hungry, ask yourself: "Am I hungry, or am I addicted?" That mindset shift changed everything.

It's the same with anything you consume.

Is it good for me, or is it harming me?

This applies to food, entertainment, and even people.

There are lessons in everything—even pain. **But some lessons aren't worth learning the hard way.**

I used to chew Tums all the time—my daughter even teased me about it. Now? I don't need them. My diet changed, my discomforts disappeared, and I threw the Tums away.

Simple changes can make a huge difference.

Instead of watching TV, get outside. My old neighbor used to be out with his kids by 8 AM, shooting hoops. That's the kind of example to follow.

And for me, that's why I stay away from the "gym rat" lifestyle—I'd rather be present with my family, doing activities together.

Sign up for something that challenges you.

It doesn't have to be extreme—a three-mile run, a monthly event, something to push you. The best medicine is movement, connection, and time outdoors.

Take care of what you consume, because it's shaping you—physically, mentally, and spiritually.

Pay attention to the fruits in your life. Are they nourishing you or poisoning you? Are they pushing you forward or holding you back?

Make the choice that serves you, your family, and your future.

8

FULFILLMENT

The Ongoing Mission

Imagine entering the military and, instead of facing roadblocks from the VA and government when it comes to your money and benefits, you're guided every step of the way.

Picture a system where, from the moment you enlist, there's no one telling you "no"—only mentors showing you how to maximize your opportunities.

What if, at 18 years old, a guiding hand was placed on your shoulder as you stepped into service?

What if they said, "From now on, you're not just earning a paycheck—you're building financial security for yourself and your family. At every opportunity, we'll make sure you know how to take full advantage of what you've earned."

Take the GI Bill, for example.

Right now, Service Members have to contribute a certain amount before they can start benefiting from it. But financial education should start much earlier.

By the time you finish basic training and arrive at your first duty station, you should already be thinking about financial independence, multiple income streams, and entrepreneurship.

That's the path to true Fulfillment—embracing the Abundance our Heavenly Father provides.

Now, imagine if Service Members were taught how to leverage their benefits wisely from day one. They'd learn how to invest VA money, use the VA home loan to buy and rent properties, and build wealth over time.

By the time they're ready to transition out of the military, they wouldn't be starting over—they'd already have financial stability and a growing portfolio.

Groups like Elite Meet exist to help Veterans, especially Special Operators, transition into civilian life. These elite warriors spend a decade "running and gunning," then suddenly find themselves lost at 28 or 30, starting from scratch.

And if they're struggling, imagine the average Service Member—how many of them are left completely unprepared?...**All of them**

That's why we need to start financial guidance at the beginning of military service, not at the end. This is what I do for my Clients.

I help them navigate the financial maze the government doesn't prepare them for.

Sadly, many only realize what they've missed when they're in their 50s or 60s, looking back at decades of missed opportunities.

One of my Clients, for example, didn't discover what was owed to him until his 65th birthday.

He received $20,000+ in back pay—then another $20,000+ the next month...Another $50,000 from another!!! Soon, he had nearly $100,000 that had been rightfully his all along.

He called me, overwhelmed, saying, "Albert, I've never had this kind of financial security. For the first time, my 10-year and 20-year plans are happening today. But damn, I wish I had known about this earlier."

That's what I hear time and time again.

Excitement, then frustration.

These men—smart, hardworking, dedicated—spent decades struggling, arguing with their spouses over money, **feeling incomplete**. And then, when they finally receive what's owed to them, they realize how much time was wasted.

One of the most powerful things my 65-year-old Client told me was this: "Albert, tell your younger Clients—**don't wait**. The old guys say: **get it done now**. **Don't live with regrets**. **Don't sit around wondering where your money is**. **Take action**."

That's the lesson here. No one ever says, "I regret taking control of my finances early." But countless people regret waiting too long.

I've never met a Service Member who was completely Fulfilled by their military career alone. They had good times, great friendships, incredible experiences—but they always thought there was another mountaintop to reach.

That's why financial Freedom matters. Because true Fulfillment comes when you have control over your future.

So whether you're 18 or 80, the time to take action is now. Fill in the gaps of your past where you can, and where you can't, learn from them and push forward.

Your money is out there. It belongs to you. But you have to take the steps to claim it.

That's why I wrote this book—to make sure no Veteran is left in the dark.

Right now, 5,000 people a day are searching for answers just like the ones in this book. And yet, when I researched, I found almost nothing out there offering real solutions fast.

Veterans are lost because they've been kept in the dark. But now **you know**.

And when you **know better**, **you do better**.

The biggest lesson?

Pay It Forward

As you take control of your own finances...help the next generation. Mentor them. Guide them. Make sure no one else has to go through the same struggles you did.

That's why we created the Five Fuck$ Nation and other online communities—so Military, Veterans, and Families can come together, support each other, and share this knowledge.

We all need each other. And the more we help each other, the stronger we become.

So push forward. Keep moving.

And as you secure your financial future, turn around and help someone else do the same.

Now, turn the page and take the next step.

III

THE FIELD MANUAL
How To Implement & Win

9

THE FIELD MANUAL AND FINAL THOUGHTS

(Knowing this isn't enough. Here's how to make sure you ACT TODAY.)

Take Action Now

Knowing this information is not enough—you have to act. It's time to make a change and take ownership of your future. I have more resources, more connections, and a community ready to support you. But **you have to take the first step**.

No More Excuses

The three most common things I hear from people are:

1. **"I don't know what I don't know."** —That's a fair excuse... **until now**.
2. **"Albert, I wish I had known this sooner."** —I get it. But you know now.
3. **"I wish I had done something sooner."** — That's the hardest one to hear. But the good news? It's not too late.

The government has kept you in the dark, but now that you know better, you have to do better. Take ownership of every day and make a move.

That's why I wrote this book. That's why I created the Five Fuck$ Nation —a network of like-minded people on Facebook, LinkedIn, and beyond who are ready to help each other succeed.

Step 1: Find The Gaps

You have to identify what's missing in your life. You're not alone in this journey—none of us were alone in the service, and we're not supposed to be alone now.

The biggest mistake people make is isolation. You need mentors who have been there, done that, and beaten down the path ahead. And in return, you need to pass that wisdom on to those behind you.

So, whether you're 18 or 80, still serving or long retired—if you're alive and breathing, you have a **responsibility** to take action.

Step 2: Flip The Script

Make the system work for you. You are in control of your life. Your money is out there—it's just locked behind bureaucratic red tape.

The difference between those who succeed and those who don't?

One fills out the paperwork, pushes forward, and claims what's rightfully theirs.

The other sits back and waits.

Be the first one. Take control. Help others take control.

We're not just individuals; we are a unit. We move together. There are 20 million of us out there.

Imagine the power if we all stood up and took what's ours.

Step 3: Build Wealth, Health, and Purpose

Don't waste years like so many before you.

Start now. Start today.

Your Fulfillment isn't just about money—it's about living with purpose.

It's about creating multiple streams of income, securing your family's future, and having the Freedom to live life on your terms.

Step 4: Lock It In

Never go back.

Take control of your Abundance, your Fulfillment, and your Future. Learn from those ahead of you. Guide those behind you. This isn't just about you—it's about an entire community rising together.

Final Thoughts: **Move Forward!**

Closing Words: **You Can't Win If You Don't Play.**

I hope this book has lit a fire under you. But now, the real work begins. Knowing better means doing better. **This is your moment—don't waste it.**

Visit my website, The Fulfilled Veteran Project.

Join the community. Become a client and get the support and guidance you need to **get paid**, and surround yourself with people who are moving forward.

The worst thing you can do is read this book and do nothing. Too many people do things like that.

But that's not you.

You wouldn't be here if you weren't ready for more. **So, get off your butt and take action today.**

Stop waiting for permission.

Your Abundance has already been granted by our Heavenly Father.

You don't need a permission slip to claim what is yours.

Your mission continues.

You are in charge.

You are in control.

Do something today.

Now, take this book, share it, buy another for a friend, and help spread the message.

Your future self will thank you for the actions you take right now.

Turn the page—**and step into your Fulfilling Future**.

Tactical Terms and Rules

I wanted to share this section with you so you understand the Tactical Terms and Rules I live by and teach every person I come across.

These aren't from the VA or any other government agency.

They're built from real experience, hard lessons, and the truth the system won't hand you.

What's inside this Tactical Terms & Rules section:

Terms — real-world definitions I created to help you see the hidden rules around money, paperwork, evil, truth, God, and the tactics that actually matter.

Rules – strategic principles to help you navigate the system, build your team, stay strong, and claim the Abundance and Fulfillment you've earned.

What it's not:

- **Not legal jargon.**
- **Not government instructions.**
- **Not academic theory.**
- **Not motivational fluff.**

You'll find what works – and the real understanding you'll need to win your battle.

Learn them. Lean on them. They're here to help you take what's already yours.

Terms

1. **Pay Charts** – The Military has clear pay charts, but the VA doesn't. The VA doesn't have a pay chart so I created one. You can see this and other true resources on my Five Fuck$ Nation groups via LinkedIn and FaceBook
2. **Good and Evil** – Our Heavenly Father is real, and evil is real, but evil doesn't make anything—it only

takes. It takes from those who don't know better and those who don't do better. This principle is important when navigating to your Fulfillment and Abundance

3. **Funnels** – In marketing, funnels help organize the sales process or knowledge process. In the Military, we focus on "Be, Know, Do." Veterans and Service Members don't know what they don't know, and that's where the confusion sets in.
4. **Be, Know, Do** – This Military mantra means you can't do better until you know better, and you can't be better until you do better. For Veterans and Service Members, this applies for your Abundance and Fulfillment
5. **A Packet** – A "packet," which is essentially paperwork. Your money is always on the other side of paperwork with the Gov't [mostly]. Whether you're filing for a fresh claim or VocRehab, etc., you need to prove your case with the required evidence, just like in Jump School or Flight School...different outcomes require different criteria.
6. **Evidence Types** – The VA requires three types of evidence:

    - **Medical Evidence** – Best from outside the VA (x-rays, reviews).
    - **Legal Evidence** – Proper legal documentation to Feed or Fight the VA
    - **Administrative Evidence** – Required paperwork from the VA.

7. **Brainwashing** – Too many Military, Veterans, and Families have been lied to by evil. The first month working with my Clients, is dedicated to removing as much brainwash as possible and letting you guys see the truth.
8. **Miranda Rights** – Unlike law enforcement, the VA doesn't have to tell you your rights to remain silent, or the right to address your attorney. Evil loves taking from you...you have Heavenly Rights so I'll show you how to use them.
9. **Trap Doors** - When evil tries to steal from you via tricks they play. Your attorney and I will call out these trap doors early so you have time to and confidence to keep moving forwards.
10. **Disability** - A lie the VA has told you for decades - the claim is for Compensation and Pension ....not Disability
11. **Compensation and Pension -** Life-changing money you've earned through your Service...not Disability
12. **Permanent and Total (P&T)** – This refers to the highest level of VA Ratings, which provides you the most money and most Freedom.
13. **Audit Proof - Not Bulletproof** - I want all of my Clients to be Audit Proof. This is doing all tasks to standard via compliance. Noone is Bulletproof when it comes to evil messing with you...just Audit Proof

14. **Human Math** – The VA uses alien calculations for rating percentages. My goal is to make sure your individual ratings add up to over 500 by the time I'm done with you.
15. **SMC S and CRSC** – Special Monthly Compensation (SMC-S) [about $400 more a month] and Combat-Related Special Compensation (CRSC) [~$2,000 a month for Retired and Medically Retired folks]
16. **eBenefits** – A platform where you need to list your family members correctly to ensure they are eligible for benefits. Too many Military, Veterans, and Families miss this step, potentially losing out on thousands in pay
17. **VocRehab and VR&E** – These are programs related to employment and rehabilitation. VR&E (Vocational Rehabilitation and Employment) provides support for those who are struggling to transition back to civilian life. [did you see the bultshit line AI spit out lol] ...I'll leave that there so you can see evil lie to your face
18. **Back Pay** – VA back pay is the money you're owed from the effective date of your claim. It can be retroactive, but it's complicated, and you need to start your back pay clock by filing your Intent to File. The VA is not the only entity that has Back Pay...like anything, this can be a tailor fit conversation at the right time for each of you.

19. **Airborne Personalities** - You all have, what I call Airborne Personalities. Some of you want to jump now, others in the middle, and those who want others to go first before you jump. I want you all to be within your comfort zone but you will feel all sorts of emotions on your Journey ...but you're not Alone
20. **Don't Overthink or Underthink** - Before you decide on anything, please talk to me so I can help you stay on track. You often overthink and believe the VA should pay you for something that happened [like a fall or surgery] but the VA doesn't do that... in such simple terms. Others, you want to Underthink...."shouldn't the VA already know this?" you ask.
21. **Drama and Trauma** - If the VA paid for Drama and Trauma, why aren't widows getting paid massive amounts of money immediately after their spouse passes?

Rules

1. **Rule of Evidence** – You must present sufficient evidence to support your claim. There are three types of evidence that are critical: medical, legal, and administrative. Each type plays a role in the success of your outcome.
2. **Rule of Knowledge** – As a Veteran or Service Member, you must learn what you don't know. Evil

will try to confuse you, and that's why it's important to have a team that can guide you through the process. You need the right knowledge to make informed decisions.

3. **The Rule of Camaraderie** – Veterans miss the camaraderie of Service, but it doesn't end there. It's about pushing forwards with your connections with others and maintaining that bond through groups like the Five Fuck$ Nation on LinkedIn and Facebook. That's the place where we can help each other. Embrace the Suck Together
4. **Rule of Team** – Evil has a system stacked against you, you need a team of good people on your side. This includes legal support, financial guidance, and Veteran advocates...just to name a few.
5. **The Rule of Abundance** – Abundance comes through our Heavenly Father [it's strange AI didn't add that so I did] knowledge, understanding, and community support. It's not about relying on the government, but about empowering yourself and your family to capitalize on the Life-Changing Money you've earned.
6. **Rule of Support** – You cannot do this alone. Surrounding yourself with knowledgeable people who care is essential. Support your family and other Veterans.
7. **The Rule of Back Pay** – You need to start your back pay clock early by filing an Intent to File. This is a critical step in ensuring you receive the retro-

active benefits you deserve. [The VA has ultimate power and say here so I agree but I have to tell you that]

8. **Rule of Responsibility** – Our Heavenly Father says you have Abundance but he isn't going to lay it in your lap. You have to join us so we can Embrace the Suck...Together
9. **The Rule of Fulfillment** – Fulfillment comes from Our Heavenly Father...you have to Bring Joy to Others
10. **The Rule of Action** – You must act quickly. Time and money cannot be replaced. The more you wait, the more you lose, so take steps now to get the Life-Changing Money you've earned.
11. **The Rule of Legacy** – Think about your legacy. It's not just about your children, but about your community, church, and the people you interact with. Help others, pay it forward, and make sure your family is properly listed for benefits.[AI should lay it on thick again about Abundance and Fulfillment and helping others....but I had to put that here]
12. **Rule of Trust** – Trust in the process and the people helping you. When you work with someone who has your best interests at heart, it becomes easier to navigate the confusing world of VA benefits. [God says to live without Fear...Give your fear to Him so you can focus on what He wants for you....Your Abundance...Your Fulfillment...You Bringing Joy to Others]

About the Author

Albert Winks, SDVOSB

As a former Army Enlisted Infantryman turned MEDEVAC Helicopter Pilot/Captain, I led the aeromedical evacuation of over 2,000 patients and I personally flew 268 combat missions evacuating over 350 patients from the battlefields of Afghanistan.

Since my Service, I've been blessed to help deliver Life-Changing Money to Military, Veterans, and their Families they've earned through their Service.

To date, I've delivered over **$10 Billion** to the **Best** and **Deadliest Humans** on **Earth**.

My Clients include MARSOC, 160th SOAR, Delta, Navy SEALs, Army Rangers, Green Berets, Top Gun pilots, Marine One crews, Blue Angels, Fighter Pilots, Air Force One and Two teams, PJs, CIA, FBI, DEA, Border Patrol, Drug Interdiction teams, Airline Pilots — and many more of America's Finest.

In my world, **Trust and Performance** aren't just words — they're the Divine Mission.

It's **Your Abundance**

It's **Your Freedom**

It's **Your Fulfillment**

It's **Your Turn**

Let's do this **Right** - **Together** - **Fast**

TheFulfilledVeteranProject.com

EPIC AUTHOR

Made in the USA
Monee, IL
03 November 2026

67380168R00075